Bruno Duarte
Arthur Santos Lopes
Lucas Lima Pereira

Brake disk facing using the DA-8700 grinding machine

Bruno Duarte
Arthur Santos Lopes
Lucas Lima Pereira

Brake disk facing using the DA-8700 grinding machine

Study of brake disk facing using the DA-8700 grinding machine at a car dealership

ScienciaScripts

Imprint

Cover image: www.ingimage.com

This book is a translation from the original published under ISBN 978-620-6-76075-7.

Publisher:
Sciencia Scripts
is a trademark of
Dodo Books Indian Ocean Ltd. and OmniScriptum S.R.L publishing group

120 High Road, East Finchley, London, N2 9ED, United Kingdom
Str. Armeneasca 28/1, office 1, Chisinau MD-2012, Republic of Moldova, Europe
Printed at: see last page
ISBN: 978-620-7-75214-0

SUMMARY

This work deals with an analysis of the machining of brake discs using the DA-8700 grinding machine, applied in the environment of a motor vehicle dealership. The aim was to analyse this type of machining with the equipment in question in the aforementioned environment, describing the conditions used, identifying the factors that led to the disc being repaired and identifying possible improvements to the machining process. By monitoring the routine at the dealership, identifying non-conformities and disc material, as well as knowing the parameters and equipment used in machining, it was possible to continue this research. In this way, it was possible to better evaluate the process and show the actions needed to improve it. The variation in disc thickness and warpage were easy to identify when carrying out the diagnosis applied at the dealership, while the identification of cutting parameters served to demonstrate that those used in machining meet the requirements of the material (FFC), even though the indexable insert used may not be the most suitable for the process. It was concluded that the use of the DA-8700 grinding machine is efficient at the dealership, carrying out the repair is an alternative to avoid premature disc replacement, and finally, in order to improve the process, it is possible to use another type of insert to reduce the need for tool replacement.

Keywords: Facing; Brake discs; DA-8700 grinding machine; Parameters.

SUMMARY

1INTRODUCTION

A vehicle's braking system must always be in good condition, as it is an important item for the safety of the vehicle's occupants. According to Limpert (1999), its main functions are to slow the vehicle down or stop it when necessary, to keep its speed under contr ol when travelling downhill and to keep the vehicle stationary when it is travelling uphill. According to Abreu (2013, p.20), the fundamental principle of drum and disc brakes is to transform kinetic energy into heat or other forms of energy, such as noise and vibration. However, drum brakes can withstand lower temperatures than disc brakes. The purpose of face turning the surface of brake discs is to correct non-conformities found on the discs, such as uneven thickness, which can cause the brake system to malfunction. It is through face turning that it is possible to obtain a flat surface (FERRARESI, 1970). This makes the use of local disc facing attractive when applied to car dealerships, due to the speed of the process.Depending on the severity of use, the temperature of brake discs can be between room temperature and 700°C, depending on the situation (MALUF et al., 2007). Because of this, grey cast iron is a viable application for ventilated brake discs for pickup trucks and SUVs, like the ones analysed in this study. Grey cast iron (CFI) has the necessary characteristics for the purpose required, due to its good thermal conductivity, good resistance to corrosion, long durability, low wear rate and good cost/benefit ratio (MALUF et al., 2007).Due to the good machinability of the material, the use of the DA-8700 grinding machine became interesting, prompting a study of it, including its cutting parameters and the cutting tool used. As well as providing a good surface finish, it makes the disc repair process practical, speeding up service at the dealership. This paper will therefore analyse this type of machining on brake discs and the importance of implementing it in a dealership.

1.1 BACKGROUND

Due to the recurring need to carry out a facing service on the front brake discs of pick-up trucks in the workshop of a dealership, we were interested in studying this procedure.

It should be noted that conventional mechanical lathes are normally used for machining, as illustrated in Figure 1.1, which are more robust than grinding machines and require greater skill in operation.

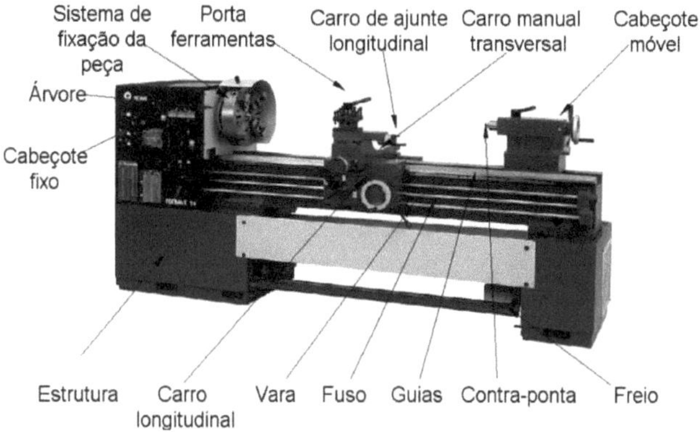

Figure 1.1 - Components of a conventional lathe *(adapted from STOETERAU, 2007).

It is also worth emphasising that by studying this procedure, it is possible to see how important it is at the dealership. This is due to the investment in training and safety of the mechanic responsible for operating the machine. Chiavenato (1999) states that training can make individuals more productive and capable of achieving organisational goals by influencing their behaviour. This makes it attractive to carry out the procedure, as it qualifies employees and eliminates the need to outsource this service.

1.2 Objectives

1.2.1 General objective

The aim of this work is to analyse, from an organisational and technical point of view, the activity of facing ventilated brake discs using the DA-8700 disc grinding machine, in a local machining process (without removing the disc from the vehicle), applied in a car dealership.

1.2.2 Specific objective

- To analyse the benefits of using this type of machining and machinery instead of conventional lathe machining at the dealership;
- Describe the operation of the DA-8700 grinding machine, the material machined and the cutting parameters applied;
- Identify the factors and causes that led to the disc being repaired;
- Describe the specifications of the disc and the conditions required to carry out the procedure;
- Identify improvements to the machining process in question.

2 LITERATURE REVIEW

This section covers the theoretical and bibliographical background to the brake system in question, cast iron, and then machining, more precisely facing, which is the focus of this study.

2.1 BRAKE SYSTEM

According to Limpert (1999), the components of the braking and steering system are extremely important in preventing critical accidents in a motorised vehicle. Canale (1989) adds that the ability to brake, whether slowing down or stopping, is an important factor in vehicle performance.

This system began to be developed at the end of the 19th century, due to the development of trains and the start of car manufacturing, thus requiring a braking mechanism (MALUF et al., 2007).

The purpose of the braking system is to stop the vehicle by the friction created between the linings when the driver presses the brake pedal (PUHN, 1987). Brossi (2002, p.28) defines its purpose as follows:classes:

The purpose of the braking system is to provide the vehicle with safe deceleration without loss of driveability and stability, in compliance with safety limits and those established by legislation. This deceleration may or may not lead to zero speed, depending on the situation imposed by the driver.

According to Limpert (1999), automotive brakes can be divided into two categories

• Drum brakes: use brake shoes that are pushed against the brake drum;

• Disc brakes: use pads that are pressed against a disc. These have now been implemented universally on the front axles of passenger cars and light lorries, as they have certain advantages over drum brakes.

As for the materials to be used in braking systems, Maluf (2007) states that they must have certain characteristics, such as good thermal conductivity,

stable friction, long durability, low wear rate, among others. He also states that there is a need to combine materials in order to fulfil the required properties.

2.1.1 Disc brake

According to Brossi (2002), disc brakes basically consist of a disc dedicated to the axis of rotation and pads applied axially against the lateral surface of the disc, compressing it. Corroborating this, Canali (2002) states that the brake pads, with their friction material, are pressed axially against the brake disc.

According to Kawaguchi (2005), this type of brake works by generating braking force through the contact of the pads on the sides of the disc due to their axial movement provided by the brake caliper piston, illustrated in Figure 2.1. Pugliesi (1976) follows a similar line of reasoning, stating that when the brake pedal is depressed and through a hydraulic process, the cylinders act on the pads, which are compressed on the disc, thus paralysing it and consequently the wheels.

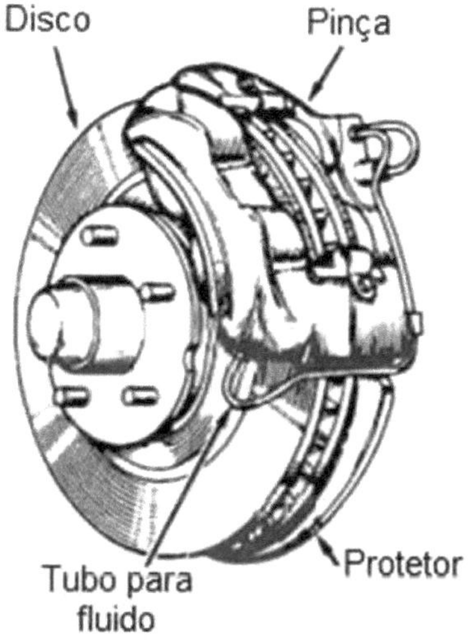

Figure 2.1 - Main components of the ventilated disc brake (Adapted from Puhn, 1987).

Some characteristics such as the geometry, size and chemical composition of the brakes influence durability, heat flow, noise characteristics and ease of maintenance, and whether they are drum or disc, both can have their performance affected (MALUF et al., 2007). According to Limpert (1999), the main advantage of using disc brakes is the low performance degradation, even at high temperatures of up to 900 °C. Brossi (2002) divides brake discs into two types, ventilated and non-ventilated (solid), stating that in non-ventilated discs there is no forced convection, which reduces energy dissipation. Ventilated discs are most commonly used on the front axle of vehicles. Kruze (2009) states that the fins inside ventilated discs generate internal air circulation due to the centripetal force when the disc rotates, which increases heat exchange through convection. It is worth noting that the ventilated disc comes in a number of variations and can be perforated and/or grooved; Figure 2.2 shows the discs mentioned.

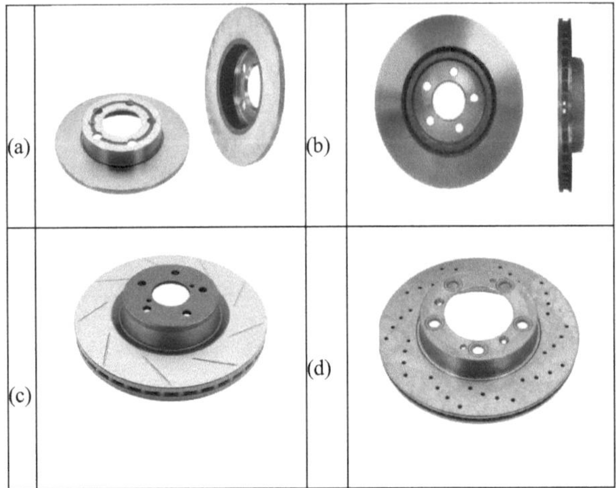

Figure 2.2 - (a) Solid brake disc, (b) Ventilated brake disc, (c) Ventilated brake disc with grooved surface and (d) Ventilated brake disc with drilled surface (Adapted from Sharp, 2011).

2.1.2 Emergence of non-conformities

According to NBR ISO 9000:2000, non-conformity is the failure to meet a requirement, and if it occurs, it may require repair, which will be an action on a non-conforming product. as can make it acceptable for use (ASSOCIAÇÃO BRASILEIRA DE NORMAS TÉCNICAS, 2000). Due to inertia, the "weight" on the front wheel axle ends up being greater compared to the rear axle wheels during the braking process, influencing braking performance due to the uneven distribution of forces on the front and rear wheels (CANALE, 1989). As a result, discs located on the vehicle's front axle are more likely to suffer damage and require repair. Depending on the road conditions, thermal history of the friction pair (disc and pads) and vehicle maintenance, variations in disc thickness can occur (CANALI, 2002). The same author also states that the high temperatures and thermal stress that discs are subjected to can cause some negative effects, such as distortion (warping) and surface cracks. Canali (2002) also states that raised points can appear on the surface of a cast iron brake disc, due to the occurrence of a large amount of heating, above the austenitisation temperature (close to 720°C), followed by rapid cooling after the friction of this region of the disc with the pad. As a result, these points become more difficult to remove, as they are more resistant to wear than the pearlite.

2.2 CAST IRON

This type of material is part of this literature review because the brake discs studied are made from grey cast iron, so it is necessary to understand the metallurgy of this product.

Machado et al. (2009) define this material as alloys that contain carbon, generally between 2% and 4%, together with other commonly used alloying elements such as silicon, manganese, phosphorus and sulphur. In addition to the above, copper, nickel, molybdenum and chromium can be found in special classes of cast iron. Ferraresi (1970) adds that the addition of nickel to cast iron increases its hardness and resistance, while the graphitising effect of nickel makes the structure easier to machine, while the addition of chromium makes the structure more resistant to wear and worsens its machinability.

Machado et al. (2009) also states that ferrite, pearlite and their mixtures are the basic structure of cast irons, adding that a FoFo structure with a matrix containing The predominance of ferrite and little pearlite makes it easier to machine at low forces, which allows for high cutting speeds and feed rates, with moderate tool wear.

Callister (2018) defines the most common types of cast iron as grey, nodular, white, malleable and vermicular. They can be characterised as follows:

• Grey cast iron: can contain up to 3% silicon, thus with a greater amount of graphite in the form of a blade, with a structure of good machinability (FERRARESI, 1970).

• White cast iron: has a low silicon content and little free graphite, resulting in a very hard and resistant structure that is very difficult to machine (FERRARESI, 1970).

• Malleable cast iron: this is annealed white cast iron. As a result of this treatment, the structure has a certain ductility and toughness, although it is easy to machine (FERRARESI, 1970).

• Nodular cast iron: this is obtained by adding a small amount of magnesium or cerium to high carbon cast iron, generating a structure with graphite in a spheroidal shape, making it resistant and as machinable as grey cast iron (FERRARESI, 1970).

• Vermicular cast iron: a recent addition to cast irons, like grey irons, it is ductile and malleable, and because of the silicon, the carbon exists as graphite (CALLISTER, 2018).

Maluf et al. (2007) states that cast irons are the most preferred materials when making brake system components, as they have good thermal conductivity and the ability to dampen vibrations, with grey cast iron alloys having the best cost/benefit ratio for use in automotive brake discs.

2.2.1 Grey Cast Iron (FFC)

The carbon content varies between 2.5% and 4% and the silicon content between 1% and 3% in grey cast irons. The graphite in this material exists in the form of flakes, surrounded by a ferrite or pearlite matrix (CALLISTER, 2018). Callister (2018) also states that due to its microstructure, the FFC is not very resistant and fragile in traction, as the ends of the graphite flakes are tapered and pointed, as shown in Figure 2.3, and can serve as points of stress concentration when receiving a traction force. in the form of veins have better thermal conductivity compared to those in the form of vermicles and nodules. Continuing this line of reasoning, Callister (2018) mentions that under compressive loads, the strength and ductility of FFCs are much greater.

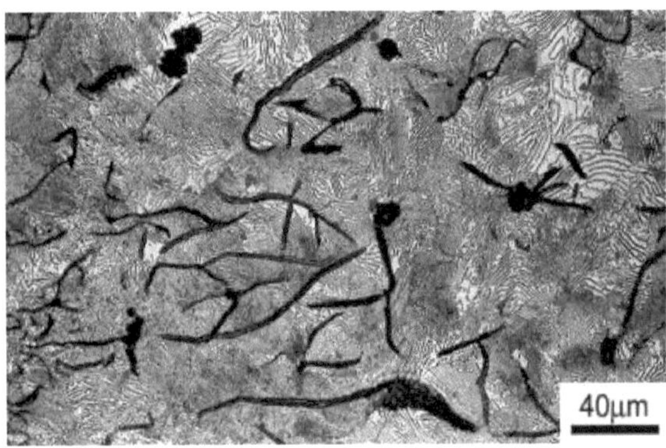

Figure 2.3 - Optical micrograph of grey cast iron containing graphite veins in a pearlitic matrix (Maluf et al., 2007).

Maluf (2007) states that grey cast iron is suitable for use in car brake discs when, in addition to high thermal diffusibility, it has good wear resistance due to the stresses it will face during braking. Cueva et al. (2001) adds that when a good relationship is achieved between thermal conductivity and mechanical resistance, it is possible to optimise a brake disc material. Specifications for critically used components in the automotive industry, such as brake discs, generally require irons with a low ferrite content, which can reduce allowable cutting speeds and feed rates, increasing machining costs and more frequent tool changes (ASM INTERNATIONAL, 1989).

2.3 MACHINING

2.3.1 General concepts

According to DIN 8580 (1985), the definition of machining applies to all manufacturing processes in which material is removed in the form of chips. Machado et al. (2009) defines it as the operation that gives shape, dimensions and finish to the part, producing chips. The chip is the part of the workpiece material that has been removed by the tool and has an irregular geometric shape (FERRARESI, 1970). Chips can basically be classified as continuous, discontinuous and segmented, with discontinuous chips being the most common type in the machining of materials such as grey cast iron, where chip breakage can occur naturally during its formation (MACHADO et al., 2009).

Parameters such as speed, feed and depth of cut affect the rate of material removed and the life of the cutting tool (STOETERAU, 2007). While Machado et al. (2009) makes an analogy between machining parameters and chip shape, stating that feed, followed by cutting speed, are the most influential in affecting chip shape.

Ferraresi (1970) defines turning as a machining process that aims to obtain surfaces of revolution with the aid of one or more single-cutting tools (which have a single exit surface on which the chip is formed), so the workpiece rotates around the main axis of rotation of the machine and the tool moves according to a trajectory coplanar to the main axis. Machado et al. (2009) list the turning operations as follows: external cylindrical turning; internal cylindrical turning; external conical turning; internal conical turning; facing; profiling; bleeding and knurling. Figure 2.4 illustrates some of these operations.

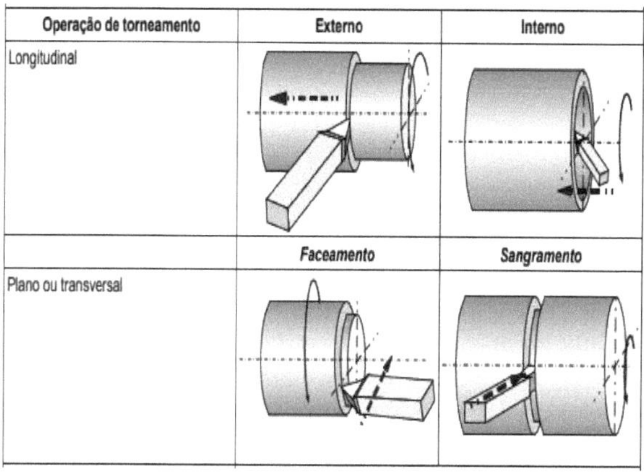

Figure 2.4 - Main turning operations (Adapted from Stoeterau, 2007).

2.3.2 Machining Grey Cast Iron

Grey cast iron has good characteristics for machining, due to its low hardness, which is generally up to 270 HB, and relative ductility. Because of these characteristics, the cutting forces are low, and even at high speeds and feeds there is economically acceptable tool wear (MACHADO et al., 2009).

Grey cast iron has good machinability, in which the presence of silicon influences its machinability, improving it. This material ends up forming break chips (discontinuous), while malleable and nodular cast irons form long chips (DINIZ, MARCONDES and COPPINI, 2014). Following this reasoning, Machado et al. (2009) states that grey cast irons are the easiest to machine, compared to nodular and malleable cast irons, where grey cast irons have short, brittle chips.

2.3.3Face turning

When radial turning aims to obtain a flat (levelled) surface, the turning is called facing (FERRARESI, 1970). In this process, the tool moves along a rectilinear path perpendicular to the machine's main axis of rotation, as shown in Figure 2.4 (BESKOW, 2011).

2.3.4Cutting tool

Machining performance depends heavily on the geometry of the cutting tool, because even if the tool is made of a good material, if its geometry is not suitable, the operation cannot be successful (MACHADO et al., 2009).

When selecting the cutting tool material, a number of factors must be taken into account, including the microstructure of the material to be machined, the type of chip, the machining process, the conditions of the machine tool, the machining conditions and the characteristics of the tool material, such as wear resistance and hardness (ALMEIDA, 2010).

Among tool materials, cemented carbide is readily available, making it quite common in machining processes. Its widespread use is due to some of the material's characteristics. According to Diniz, Marcondes and Coppini (2014), cemented carbide has high hardness and resistance to compression, and is most often used in the form of inserts that are mechanically fixed (interchangeable) on a tool holder.

3 METHODOLOGY

This section discusses the methodology used in the research, containing information on how discs are faced. This study aims to analyse the process of brake disc facing using the DA-8700 grinding machine at a motor vehicle dealership. The methodology adopted included careful selection of the dealership, followed by data collection through observation of the work routine, identification of non-conformities in the discs, and analysis of the machining parameters used. Through the diagnosis applied, it was possible to identify variations in disc thickness and warpage, as well as determine the efficiency of the machine in carrying out the repair. In addition, the conformity of the cutting parameters with the material requirements of the discs was assessed.

In addition, the study highlights the importance of preventive maintenance of brake discs, emphasising the importance of early diagnosis of non-conformities. The analyses carried out provided a deeper understanding of the challenges faced in the machining process, allowing opportunities for improvement to be identified. Finally, the conclusions reached provide a solid basis for implementing improvements in the working environment, promoting a more efficient and reliable operation. This study thus contributes to continued progress in the field of automotive maintenance and the optimisation of brake disc machining processes in dealerships

3.1 APPLICATION OF FACING AT THE DEALERSHIP

Initially, the application of facing comes from a customer's claim when they take their vehicle to the dealership for maintenance. Usually, the complaint is about vibration when the brake pedal is pressed, so it is necessary to follow the flowchart in Figure 3.1.

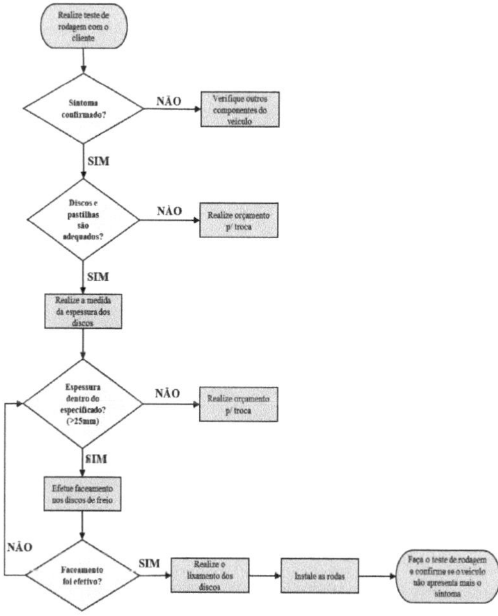

Figure 3.1 - Flowchart for diagnosing and carrying out facing (Author, 2021).

Thus, the mechanic responsible for carrying out the service must follow the steps in the flowchart, carrying out the disc repair or action needed to solve the car's problem. 8700 GRINDING MACHINE

3.1.1 General information

The purpose of the machine in question is to perform face turning on brake discs, making their surface flat, thus eliminating the variation in thickness and warping that the disc may have suffered. It is installed in the vehicle itself and its positioning is shown in Figure 3.2.

Figure 3.2 - Ventilated brake disc facing process (Author).

The machine is easy to handle and move around, allowing it to be moved around the workshop, but it is not very robust when compared to a conventional lathe. It is divided into a grinding head (DU8704) and a control unit (DU8610), as shown in Figure 3.3.

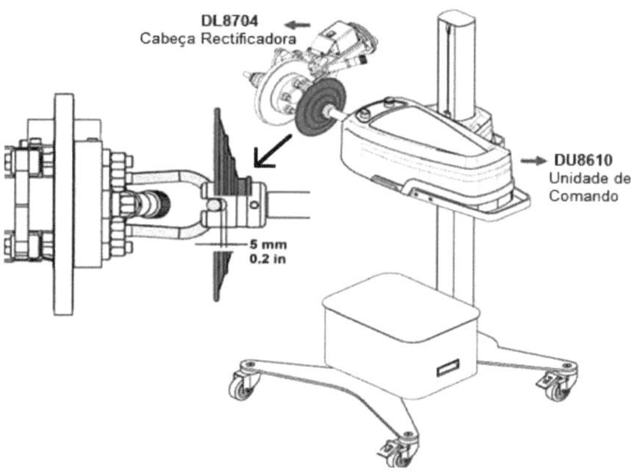

Figure 3.3 - DA8700 Disc Grinder and its components, DL8704 Grinding Head and DU8610 Control Unit (Adapted from MAD HOLDING B.V., 2015)

The grinding head is the component responsible for the feed and depth of cut applied to the turned part. Where the cutting depth is controlled by the cutting depth adjustment knob and the feed by the automatic distribution switch, without the need for interference from the equipment operator, Figure 3.4 details the grinding head controls.

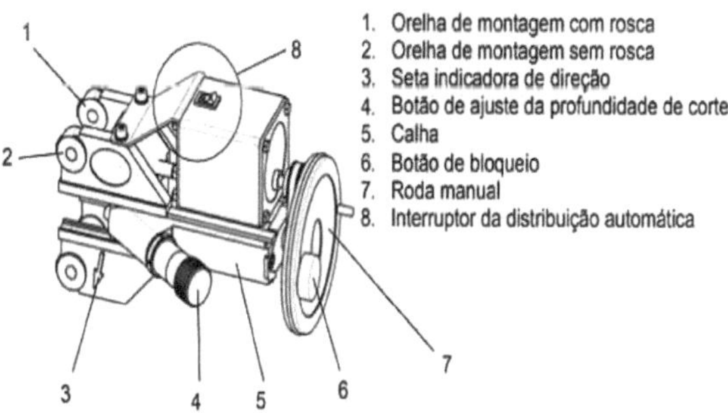

Figure 3.4 - Grinding head controls (Adapted from Mad Holding B.V., 2015).

While the Control Unit is the component responsible for rotating the part, in this case the ventilated brake discs, as well as directing the direction of rotation and promoting an emergency stop if necessary. The aforementioned controls can be seen in Figure 3.5.

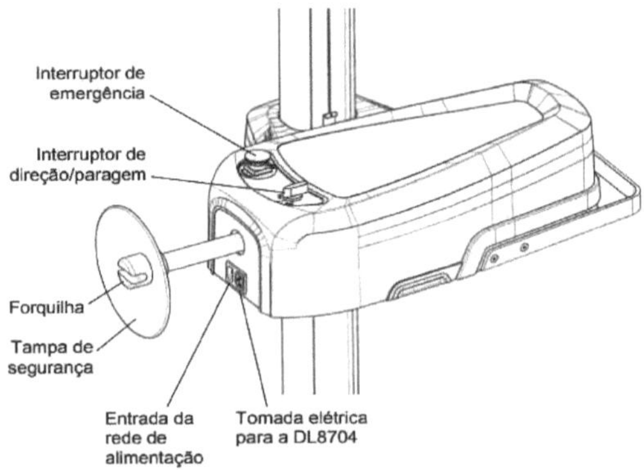

Figure 3.5 - DU8610 control unit controls (Adapted from MAD HOLDING B.V., 2015).

3.1.2 Cutting parameters

When turning, certain cutting parameters must be observed, such as feed rate, feed speed, workpiece rotation and cutting speed. Table 3.1 provides data on the parameters used in turning.

Table 3.1 - Parameters used in facing

Turning cutting parameters	
Progress	0.085 mm/rev
Feed speed	8.5 mm/min
Rotation	100 rpm
Cutting speed	100 m/min (approximately)
Adjustment knob cutting depth accuracy	0.05 mm

In addition to the parameters mentioned, it is important to emphasise the insert used for turning. The insert in question is the TPGT 110204 FN AL H10 carbide, shown in Figure 3.6.

Figure 3.6 - TPGT 110204 FN AL H10 insert (Author, 2021)

As this is a machining process, it is important to show the classification of the insert used according to ABNT NBR ISO 1832. Table 3.2 deals with this classification.

Table 3.2 - ISO coding for inserts

Parameters		Specification
T	Insert geometry	Triangular - angles of 60°
P	Clearance angle	11°
G	Manufacturing tolerance	±0.025 diameter, ±0.025 height and ±0.13 thickness
T	Clamping geometry and chip breaking	Through-hole and single-sided profile
11	Wafer edge size	6.35mm (1 ") diameter of the inscribed circle4
02	Wafer thickness	2.38 mm
04	Wafer radius	0.4 mm
F	Edge preparation	Pointed
N	Forward direction	Neutral cut
L	Chip breaker application	Light cut

3.1.3 Carrying out the repair

Once the symptom of a problem with the vehicle's brake disc has been confirmed, following the flowchart in Figure 3.1, the disc is repaired. This consists of several passes to a depth of 0.05 mm (applied equally to both sides of the disc), until a visual inspection shows that the disc has been fully recovered. First of all, the Control Unit must start rotating, following the direction of the direction indicator arrow shown in Figure 3.4, followed by setting the cutting depth and starting the automatic feed (in the direction shown in Figure 3.7) carried out by the Grinding Head. In addition, the indexable inserts must be correctly positioned, as also shown in Figure 3.7.

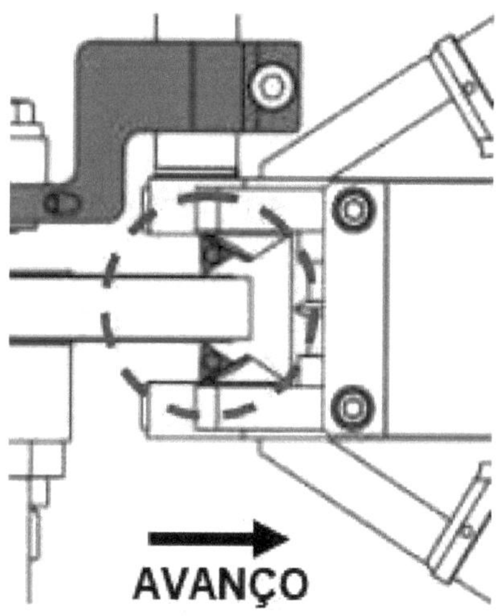

Figure 3.7 - Indication of feed and correct positioning of the inserts during facing (Adapted from MAD HOLDING B.V., 2015).

3.2 INSTRUMENTATION

To carry out the facing process, it is necessary to use precision instruments to measure certain parameters, such as torque and dimensions. Because of the importance of these instruments, those needed to carry out the operation will be described in more detail.

3.2.1 External micrometer

It is an instrument that provides extreme precision, used for measuring different external surfaces (TRAMONTINA, 2017). Because it is used for rigorous measurements, with an accuracy of 0.01mm, it is suitable for measuring the thickness of discs. From this it is possible to determine whether the disc still has the minimum safe thickness indicated by the manufacturer. Therefore, it is appropriate to use a micrometer with a capacity of 25mm to 50mm and a resolution of 0.01, similar to the one in Figure 3.8, because the minimum thickness required for facing is greater than 25mm, and if the disc has a thickness of less than 25mm, it becomes unusable.

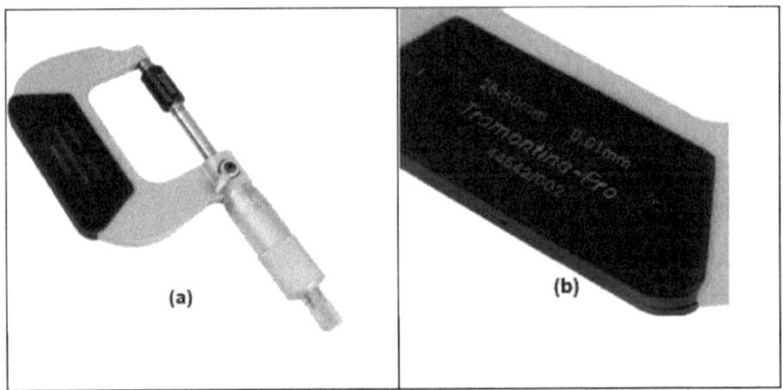

Figure 3.8 - (a) Tramontina external micrometer; (b) capacity and resolution detail (Adapted from GurgelMix Máquinas e Ferramentas S.A, 2021).

In order to guarantee the safety of the disc thickness measurements, it is necessary to take measurements at 12 points on the brake disc at 10mm from the outer edge, as shown in Figure 3.9, making it possible to check the variation in disc thickness.

Figure 3.9 - Using a micrometer to measure brake disc thickness (Author, 2021).

3.2.2 Snap torque wrench

Torque wrenches are used to check the torque when tightening bolts or nuts, because if the torque is less than specified, the bolt can loosen over time, However, if the tightening is greater than specified, damage to the thread fillets may occur (TRAMONTINA, 2017). This precision tool is used to properly torque nuts and bolts, such as those that hold the brake disc and the Grinding Head (DL8704) in the vehicle, which require a torque of 50 N.m. It is therefore necessary to use a good precision torque wrench with a resolution of 1 N.m, similar to the one in Figure 3.10.

Figure 3.10 - 40 - 200 N.m snap torque wrench, 1 N.m resolution (GurgelMix Máquinas e Ferramentas S.A, 2021).

3.2.3 Dial indicator

It is a very sensitive precision instrument, used to check measurements, flat surfaces, concentricity and parallelism, and can be used to measure or compare the shapes of parts (SENAI, 1996).The dial indicator is used to check for warping in the vehicle's wheel discs and hub, as shown in Figure 3.11.

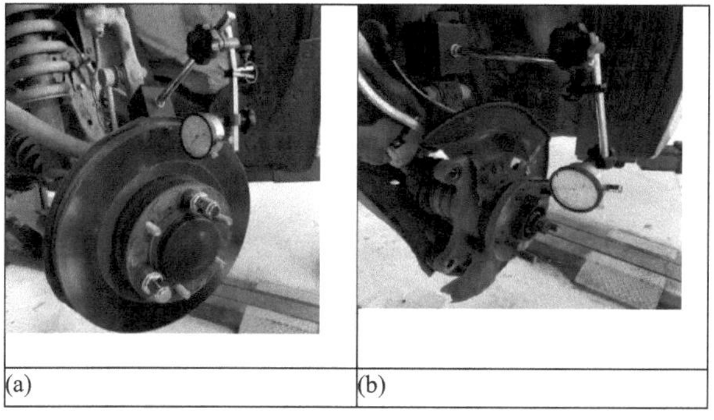

Figure 3.11 - Use of the dial indicator on: (a) disc; (b) wheel hub (Author, 2021).

3.3 BRAKE DISCS

The ventilated discs used are of similar configuration to those manufactured by TRW, made of grey cast iron, with 3.45% carbon, high wear resistance and low melting temperature, between 1140°C and 1250°C (ZF FRIEDRICHSHAFEN, 2020). The material used in the discs is similar to the G3000 designation (G10H18 and G11H18) in the SAE J431 standard, due to its similar chemical composition to those shown in APPENDIX A and its application in brake discs, which APPENDIX B describes. In addition, according to the same standard, contained in ANNEX C, the matrix of its microstructure is pearlitic. It is worth noting that due to the presence of fins, the disc has a high thermal capacity, as they provide air circulation inside. The disc used is more detailed in Figure 3.12, where you can see its dimensions.

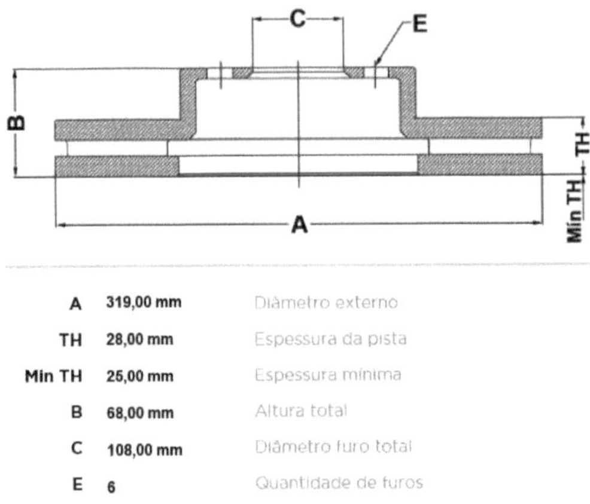

Figure 3.12 - Dimensions of the brake disc used (Adapted from Destak Autopeças, 2021).

3.4 VISUAL ANALYSIS OF THE DISC

Visual analysis is extremely important, as it allows you to see what action the heat may have had on the disc, changing its surface and whether the disc repair was successful. Figure 3.13 shows a brake disc that has been used a lot and another that hasn't been used yet, with a very different surface, the older disc having suffered the effects of temperature and friction as a result of use.

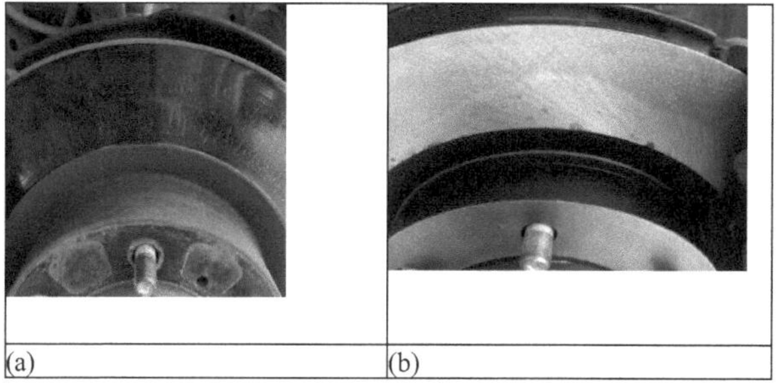

(a) (b)

Figure 3.13 - Brake disc: (a) very worn; (b) new (Author, 2021)

Visual analysis is also used in the facing process because of the change in surface appearance; at the end of facing, the brake disc should have a visually different surface finish. Figure 3.14 shows a disc during facing, with a noticeable difference in surface finish. In this case, it was not possible to completely repair the disc in a single cutting operation, requiring repetition with an increased depth of cut, which could make the surface homogeneous.

Figure 3.14 - Facing disc (Author, 2021).

4 RESULTS AND DISCUSSIONS

The execution of this work required a study of brake disc facing, specifically applied to a car dealership, requiring observations on machinery, brake discs and employee routines. It was therefore necessary to determine a number of stages for carrying out the study, shown in Figure 4.1.

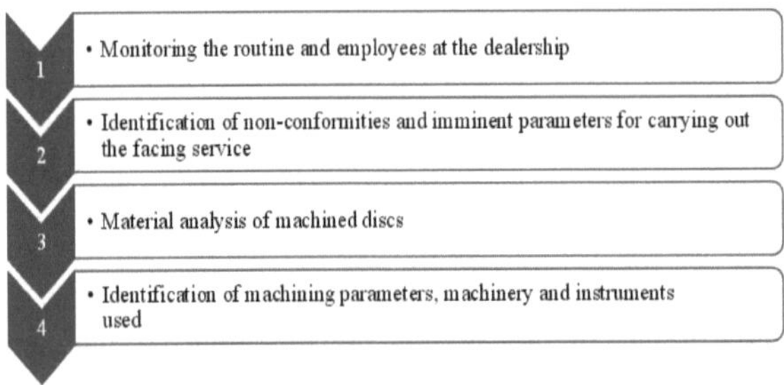

Figure 4.1 - Steps that led to the study (Author, 2021)

Carrying out the steps described in Figure 4.1 helped identify many aspects of the braking activity. Firstly, it was possible to see how the workflow works, so that the vehicle's brake system could be diagnosed and, if necessary, repaired. Continuing this line of thought, it was possible to observe the presence of non-conformities in the discs using precision instruments. For example, the variation in thickness, cited by Canali (2002) as something that can happen with discs, where by measuring with a micrometer it was possible to check the variation in thickness, and also to check that the disc is at a safe thickness for the job. The use of a dial indicator is used to check for warping in the part. Based on information from the manufacturer and observation of standards such as SAE J431, shown in the Annexes to this work, it was possible to better characterise the material and composition brake discs. With the use of grey cast iron with a pearlitic matrix, it has become necessary to observe the machinability and mechanical strength of the material, linked to its application. After all, according to Chiaverini (1988), when ferrite predominates in the matrix of cast irons, the machinability of the material is better, but its mechanical strength and resistance to wear are impaired, however, if pearlite is the predominant constituent in the matrix, the material will have better mechanical strength.

Once the machinery and machining parameters had been identified, it was possible to check that they met the needs of the material being machined. After all, brake discs generally have a low ferrite content in their composition, which can make machining difficult and tool changes more frequent (ASM INTERNATIONAL, 1989). Therefore, during the facing process, a cutting speed of 100 m/min is used, as shown in Table 3.1, which is not so high when compared to the cutting speeds indicated for stainless steel, for example, which according to Machado et al. (2009) can be up to 400 m/min.It is important to note that the indexable insert used, TPGT 110204 FN AL H10, may not be the most suitable for the process, as the indications on its label do not list cast iron as a suitable material, and there is no "Cast Iron" option ticked, as shown in Figure 3.6. However, the insert does have indications for materials such as stainless steel, which can be machined at much higher cutting speeds than cast iron.In order to match the cutting tool to the cutting parameters and machined material, it is important to look for options available on the market. Among the options available, the TNMG 160408 HA insert could be a good alternative, as it has cast iron as the material indicated for use, as shown in Figure 4.2, with "K" (cast iron) on its label and a cutting speed of between 100 and 200 m/min, similar to that used in the disc repair process.

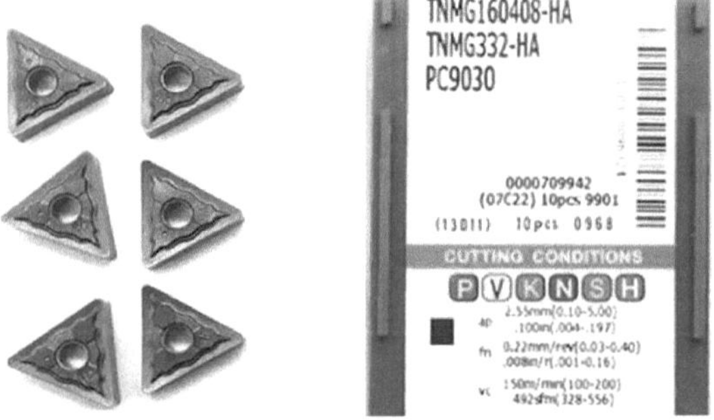

Figure 4.2 - TNMG 160408 HA insert and specifications (Adapted from Suprir Ferramentas, 2021)

5 CONCLUSION

In view of the above, and bearing in mind that this work aimed to analyse face machining using unusual machinery in a car dealership, the considerations made above point to the following conclusions:

- Facing using the DA-8700 grinding machine proved to be efficient because it added speed to the disc repair process and did not require outsourcing the service (as it was not carried out on a conventional lathe outside the dealership);
- Because the DA-8700 grinder is easy to handle and maintain, training employees in its operation can bring a quick return for the dealership;
- This type of machining has become an alternative to avoid premature replacement of brake discs, which could lead to higher costs for customers;
- In order to improve the process, it is necessary to use another type of indexable insert, such as the TNMG 160408 HA insert, because it can better adapt to the machining parameters used and reduces the need for tool changes.

5.1 RECOMMENDATIONS FOR FUTURE WORK

In view of the above, some subjects that are considered important to explore can be used in later work:

- Facing of ventilated brake discs on the DA-8700 Grinding Machine using cutting fluids;
- Study of the wear of interchangeable pads using cooling fluids during the facing of brake discs;
- Study of the feasibility o f applying heat treatment to brake discs after facing;
- Root cause analysis of the problem of warping ventilated brake discs.

REFERENCES

ABREU, R. M. D. **SIMULATION AND TESTING OF A BRAKE MECHANISM AUTOMOTIVE**. Dissertation (Master's Thesis) - UFMG. Belo Horizonte, p. 152. 2013.

ALMEIDA, C. M. D. WEAR **EVALUATION OF HARD METAL TOOLS COATED WITH TiN IN THE FRAMING OF LAMINATED ABNT4140 STEEL HOT AND TEMPERATE/COOL**. Dissertation (Master's Thesis) - PUCMG. Belo Horizonte. 2010.

ASM INTERNATIONAL. **Metals Handbook:** Machining. 9th ed. Ohio: American Society for Metals, v. 16, 1989.

BRAZILIAN ASSOCIATION OF TECHNICAL STANDARDS. **NBR ISO 9000: Systems of Quality management - Fundamentals and vocabulary**. Rio de Janeiro, p. 26. 2000.

BRAZILIAN ASSOCIATION OF TECHNICAL STANDARDS. **NBR ISO 1832:2016: Tablets for cutting tools - Designation**. Rio de Janeiro. 2016.

BESKOW, A. B. Wordpress. **MACHINING PROCESSES I**, 2011. Available at:<https://plant3d.files.wordpress.com/2011/04/processos_de_usinagem_i_-_aula_02_-_conventional_machining_processes.pdf>. Accessed on: 01 June

2021.

BROSSI, A. A. **Study of the braking performance of a bi-articulated bus**. Dissertation (Master's Thesis) - USP. São Carlos. 2002.

CALLISTER, W. D. J. **Materials science and engineering:** an introduction. 9th ed. Rio de Janeiro: LTC, 2018.

CANALE, A. C. **Automobilistica:** dinâmica, desempenho. São Paulo: Érica, 1989.

CANALI, R. J. **DETERMINATION OF THE PHYSICAL PROPERTIES OF DIFFERENT MATERIALS FOR BRAKE DISCS AND PADS AND THEIR RELATIONSHIP WITH NOISE PROPERTIES. Dissertation (Master's Thesis) - UFRGS. Porto Alegre, p. 132. 2002.**

CHIAVENATO, I. **Treinamento e Desenvolvimento de Recursos Humanos:** como incrementar talentos na empresa. 4th ed. São Paulo: Atlas, 1999.

CHIAVERINI, V. **Steels and Cast Irons**. 6th Ed. São Paulo: ABM, 1988.

CUEVA, G. et al. **Wear of Cast Irons Used in Automotive Brake Discs**. ABCM. Uberlândia. 2001.

DESTAK AUTO PARTS. Destak Auto Peças website. **Ventilated Front Brake Disc – Fremax - BD5618 - Pair**, 2021. Available at: <https://www.destakautopecas.com.br/automotivo/autopecas/disco-de-freio/disco-de-freio- front-ventilated-fremax-bd5618-par>. Accessed on: 3 June 2021.

DIN. **DIN 8580 - Fertingunsverfahren**. Beuth Verlag. Berlin. 1985.

DINIZ, A. E.; MARCONDES, F. C.; COPPINI, N. L. **Tecnologia da machinagem dos materiais**. 9th ed. São Paulo: Artliber, 2014.

FERRARESI, D. **Fundamento de Usinagem dos Metais**. São Paulo: Blucher, 1970.

GURGELMIX MÁQUINAS E FERRAMENTAS S.A. Mechanic's Shop. **External Micrometer 25 - 50 mm - TRAMONTINA PRO**-44542002, 2021. Available at: <https://www.lojadomecanico.com.br/produto/90789/3/204/micrometro-externo-2550-mm-tramontina-pro-44542002>. Accessed on: 2 June 2021.

KAWAGUCHI, H. **Comparison of subjective vs. objective braking comfort analysis of a passenger car**. Dissertation (Master's Thesis) - USP. São Paulo. 2005.

KRUZE, G. A. S. **Evaluation of the coefficient of friction in a reduced-scale dynamometer**. Dissertation (Master's Thesis) - UFRGS. Porto Alegre, p. 79. 2009.

LIMPERT, R. **Brake Design and Safety**. 2nd ed: SAE, 1999.

MACHADO, Á. R. et al. **Teoria da machinagem dos materiais**. 1st ed. São Paulo: Blucher, 2009.

MAD HOLDING B.V. **DA8700: User Manual**. MAD Disc Aligner. Veenendaal, p.36. 2015.

MALUF, O. **Thermomechanical fatigue in grey cast iron alloys for automotive brake discs**. Thesis (Doctoral dissertation) - USP. São Carlos. 2007.

MALUF, O. et al. **Automotive brake discs: historical and technological aspects**. [S.l.]. 2007.

PUGLIESI, M. **Complete Automobile Manual**. [S.l.]: HEMUS, 1976. PUHN, F. **Brake Handbook**. 1st ed. New York: HpBooks, 1987.

SENAI. **Mechanics Basic Metrology**. SENAI - ES. Vitória, 1996.

SHARP,B. **FADING, O ENEMY THE MOTORIST'S ENEMY**, 2011. Available at: <http://www.autoentusiastasclassic.com.br/2011/06/fading-o-inimigo-do-motorista.html>. Accessed on: 31 May 2021.

SOCIETY OF AUTOMOTIVE ENGINEERS. **SAE J431 Surface Vehicle Standard**. SAE International. Warrendale, 2000.

STOETERAU, R. L. Polytechnic School of the University of São Paulo. **Machining with of Geometry Defined**, 2007. Available at: <http://sites.poli.usp.br/d/pmr2202/arquivos/aulas/PMR2202-AULA%20RS2.pdf>.Accessed on: 1 June 2021.

SUPPLY TOOLS. **INSERT TNMG 160408 HA PC9030 - KORLOY**, 2021. Available at: <https://www.suprirferramentas.com.br/inserto-pastilha- tnmg-160408-ha-pc9030-cx-10-pcas>. Accessed on: 1 July 2021.

TRAMONTINA. **Tramontina PRO Catalogue**. TRAMONTINA GARIBALDI SA IND MET.Garibaldi, p. 600. 2017.

ZF FRIEDRICHSHAFEN. Discs discs. **TRW**,2020.Available at: <https://www.trwaftermarket.com/br/carros-de-passeio/freios/Discos-de-freio/>. Accessed on: 10 March 2021.

ANNEXES

ANNEX A - SAE J431 STANDARD (CHEMICAL COMPOSITION)

A.2 Chemical Composition

A.2.1 Typical base composition ranges generally employed for the iron grades are shown in Table A1. The base composition does not include alloys such as Cu, Cr, Mo, Ni, or others which may be added for hardness or t/h control, or to meet mandatory composition limits of special irons given in Table 3 of the main body of this document.

TABLE A1—TYPICAL BASE COMPOSITIONS

Iron Grade	Previous Designation	Carbon	Silicon	Manganese	Sulfur Max.	Phosphorus max	C. E.[1] (Approx.)
G7	G1800h	3.50 - 3.70	2.30 - 2.80	0.60 - 0.90	0.14	0.25	4.35 - 4.55
G9	G2500	3.40 - 3.65	2.10 - 2.50	0.60 - 0.90	0.12	0.25	4.15 - 4.40
G10	G3000	3.35 - 3.60	1.90 - 2.30	0.60 - 0.90	0.12	0.20	4.05 - 4.30
G11	G3000	3.30 - 3.55	1.90 - 2.20	0.60 - 0.90	0.12	0.10	4.00 - 4.25
G12	G3500	3.25 - 3.50	1.90 - 2.20	0.60 - 0.90	0.12	0.10	3.95 - 4.20
G13	G4000	3.15 - 3.40	1.80 - 2.10	0.70 - 1.00	0.12	0.08	3.80 - 4.05

1. C. E. (Carbon Equivalent) = %C + (1/3) %Si.

A.2.2 Typical base composition ranges may vary for specific grades depending on casting section size or metallurgical factors such as trace element content, or to satisfy mandatory composition requirements of special irons as given in Table 3.

A.2.3 Typical composition ranges including typical alloy content for camshaft iron, grade G11H24d, are shown in Table A2.

TABLE A2—TYPICAL CHEMICAL COMPOSITION OF ALLOY GRAY IRON AUTOMOTIVE CAMSHAFTS, GRADE G11H24d (PREVIOUS 4000d)

Constituent	Wt %
Total Carbon	3.10 to 3.60
Silicon	1.95 to 2.40
Manganese	0.60 to 0.90
Phosphorus	0.10 max
Sulfur	0.15 max
Chromium	0.85 to 1.50
Molybdenum	0.40 to 0.60
Nickel	0.20 to 0.45
Copper	Residual

A.3 Microstructure

A.3.1 The as-cast microstructure of gray iron covered by this document consists of a mixture of flake graphite in a matrix consisting of ferrite, ferrite and pearlite, or pearlite, as described in Table A3. The quantity of flake graphite and size of the flakes vary with iron grade. The amount and fineness of pearlite vary with the hardness grade. The pearlite is usually lamellar but may be partially spheroidal in slowly cooled sections or where heat treatment has been applied.

-12-

ANNEX B - SAE J431 STANDARD (TYPE OF MICROSTRUCTURE)

SAE J431 Revised DEC2000

TABLE A3—TYPICAL MICROSTRUCTURES OF REFERENCE GRADES

SAE Casting Grade	Previous Designation	Microstructure Graphite[1]	Microstructure Matrix
G9H12	G1800	Type VII A & B	Ferritic - Pearlitic
G9H17	G2500	Type VII A & B	Pearlitic - Ferritic
G10H18	G3000	Type VII A	Pearlitic
G11H18	G3000	Type VII A	Pearlitic
G11H20	G3500	Type VII A	Pearlitic
G12H21	G4000	Type VII A	Pearlitic
G13H19	G4000	Type VII A	Pearlitic
G7H16 c	G1800 h	Type VII A, B, & C size 1-3	Lamellar Pearlite
G9H17 a	G2500 a	Type VII A size 2-4	Lamellar Pearlite
G10H21 c	G3500 c	Type VII A size 3-5	Lamellar Pearlite
G11H20 b	G3500 b	Type VII A size 3-5	Lamellar Pearlite
G11H24 d	G4000 d	Type VII A & E size 4-7(1)	Pearlitic - Carbidic[2]

1. See ASTM A247.
2. In cam nose. As cast. matrix pearlite in cam may be transformed to tempered Martensite by subsequent Flame or induction hardening.

A.3.2 The size and distribution of graphite flakes in gray iron depend upon chemistry, liquid metal treatment (inoculation), and cooling rate during solidification. The primary, but not sole, chemical determinant is carbon equivalent, defined as C+Si/3.

A.3.2.1 Alloying elements used for pearlite hardness control have small but non-negligible effects on graphite size. Since some elements operate as coarsening and others as refining agents, combinations can be used for a neutral effect.

A.3.2.2 When alloying elements are used to produce a mixed structure of primary carbide and graphite, as in the cams of alloy hardenable gray iron automotive camshafts, eutectic graphite is reduced and significant flake refinement results.

A.3.2.3 The graphite microstructure of gray iron cannot be changed by heat treatment.

A.3.3 Hardness of the ferrite in the gray iron matrix is unaffected by cooling rate but is affected by alloy elements in solid solution, the most noticeable being silicon, which increases ferrite hardness about 35 HB for each 1% of Silicon present. Heat treatment is required to decompose all pearlite and produce a fully ferritic structure.

A.3.4 The amount and hardness of pearlite depend jointly on cooling rate and alloy chemistry, which are balanced in the foundry to control pearlite amount and hardness and, consequentially, casting hardness. Both the amount and hardness of pearlite can be altered by heat treatment.

A.3.5 In special cases such as alloy hardenable iron camshafts, alloy is also used to obtain controlled percentages of carbides, detracting from graphite, in cam and valve lifter surfaces where maximum contact stress occurs. The as-cast matrix structure in these cases is pearlite; in the contact surfaces, the matrix is transformed to tempered martensite by surface heat treatment.

A.3.6 Gray iron castings can be through-hardened by liquid quenching or selectively surface-hardened by either flame or induction methods.

-13-

ANNEX C - SAE J431 STANDARD (GREY CAST IRON APPLICATION)

A.5 Application of Gray Iron Castings

A.5.1 Typical applications of both the regular and special reference grades given in Table 4 are shown in Table A6. Iron grade combinations considered standard are not limited to these reference grades.

TABLE A6—TYPICAL AUTOMOTIVE APPLICATIONS OF GRAY IRON REFERENCE GRADES

SAE Grade	Previous Designation	General Data
G9H12	G1800	Miscellaneous soft castings (as cast or annealed) with relatively low strength requirements. Exhaust manifolds, alloyed or unalloyed. Annealed as necessary to prevent growth cracking in service due to heat.
G9H17	G2500	Small cylinder blocks, cylinder heads, air cooled cylinders, pistons, clutchplates, oil pump bodies, transmission cases, gear boxes, clutch housings, light-duty brake drums.
G10H18	G3000	Passenger car and light-duty truck cylinder blocks and heads, flywheels, differential carrier castings, pistons, medium-duty brake drums and discs, clutch plates, hydraulic castings, and refrigerant compressor castings.
G11H18	G3000	Same general uses as G10H18 with suitability for larger section sizes or where tensile strength or duty requirements are higher.
G11H20	G3500	Medium and heavy-duty truck and tractor diesel cylinder blocks and heads, heavy flywheels, transmission cases, axle housings, and miscellaneous heavy gear boxes.
G12H21	G4000	Extra heavy-duty diesel engine cylinder heads, liners, and pistons.
G13H19	G4000	Large heavy-duty diesel engine and construction equipment castings. Heavy-duty hydraulic castings.
G7H16 c	G1800 h	Brake drums and discs where very high damping capacity is required.
G9H17 a	G2500 a	Brake drums and clutch plates for moderate service requirements and where high carbon iron is desired to minimize heat checking.
G10H21 c	G3500 c	Extra heavy-duty service brake drums.
G11H20 b	G3500 b	Brake drums and clutch plates for heavy-duty service where high carbon and high hardness are both required to minimize heat checking and provide higher strength.
G11H24 d	G4000 d	Alloy hardenable iron automotive engine camshafts.

A.5.2 The castability, thermal conductivity, and vibration damping capacity of gray iron all relate closely and directly with its largely graphitic carbon content and, therefore, vary inversely with iron grade as numerically defined in this document. Machinability and wear resistance are more influenced by hardness which is controlled more by matrix hardness than by graphite content. Tensile strength is directly proportional with both iron and hardness grades and can be regulated with either one. Since castability is constant with iron grade, it is often more practical to obtain a needed strength increment by using alloy ladle additions to change the hardness grade rather than to change the iron grade. Iron grade can also be changed by ladle alloy additions which change the size of graphite flakes rather than the quantity. This has an intermediate effect on castability that is often more tolerable and therefore less limiting of design freedom than changing iron grade by changing carbon equivalent.

A.5.3 With grade differentiation by t/h ratio comparative fatigue behavior of different grades of gray iron differs with the type of loading; i.e., whether strain limited or stress limited. In the case of strain limited cyclic loading, fatigue life ("strain life," in this case) tends to be mainly controlled by matrix strength and ductility and shows little variation with t/h ratio. This means that changing iron grade as defined in this document without altering matrix metallurgy, for example by changing only carbon equivalent, will not significantly change strain life even though there is substantial change in t/h ratio and tensile strength. This behavior is shown in reference 8 (see 2.2). For this reason, when service loads are strain limited, the lower t/h iron grades are often more optimum because of their better castability and also because, with a given matrix hardness, the lower t/h grades have a lower ratio of bulk hardness to matrix hardness and hence better machinability at a given strain life. The usual way of increasing matrix hardness -- by pearlite refining alloy additions -- also increases matrix ductility. Examples of strain limited loads are temperature gradient induced loads as in brake drums and water jacketed combustion chambers, and secondary loads carried by gray iron parts incorporated in structures with stronger

Printed by Books on Demand GmbH, Norderstedt / Germany